A Decision Support System for Evaluating Ranges and Airspace

Albert A. Robbert   Manuel Carrillo
Robert Kerchner   William A. Williams

Prepared for the United States Air Force
Project AIR FORCE
RAND

The research reported here was sponsored by the United States Air Force under Contract F49642-96-C-0001. Further information may be obtained from the Strategic Planning Division, Directorate of Plans, Hq USAF.

Library of Congress Cataloging-in-Publication Data

A decision support system for evaluating ranges and airspace / Albert A. Robbert ... [et al.].
p. cm.
"MR-1286/1-AF."
ISBN 0-8330-2935-5
1. Air bases—United States. 2. Military reservations—United States. 3. Airspace (Law)—United States. I. Robbert, Albert A., 1944–

UG634.49 .D43 2001
358.4'17'0973—dc21

00-067355

The cover was prepared by Tanya Maiboroda using an image supplied by Kent Bingham, Photo/Graphic Imaging Center, Hill Air Force Base, Utah.

Published 2001 by RAND
1700 Main Street, P.O. Box 2138, Santa Monica, CA 90407-2138
1200 South Hayes Street, Arlington, VA 22202-5050
RAND URL: http://www.rand.org/
To order RAND documents or to obtain additional information, contact Distribution Services: Telephone: (310) 451-7002;
Fax: (310) 451-6915; Internet: order@rand.org

PREFACE

Officials responsible for range and airspace management at Headquarters Air Combat Command (ACC) asked RAND's Project AIR FORCE to undertake a study that would improve the collection, evaluation, analysis, and presentation of the information needed to link training requirements to their associated airspace and range infrastructure requirements and to evaluate the existing infrastructure. This study was conducted initially in Project AIR FORCE's Resource Management Program. The work shifted to the Manpower, Personnel, and Training Program when it was formed in 1999.

This report provides information on the construction, use, and maintenance of a decision support system (DSS) assembled by RAND for this project. A companion volume (*Relating Ranges and Airspace to Air Combat Command Missions and Training*, MR-1286-AF) provides findings, developed through use of the DSS, regarding the adequacy of ACC's range and airspace infrastructure.

PROJECT AIR FORCE

Project AIR FORCE, a division of RAND, is the Air Force federally funded research and development center (FFRDC) for studies and analyses. It provides the Air Force with independent analyses of policy alternatives affecting the development, employment, combat readiness, and support of current and future aerospace forces. Research is performed in four programs: Aerospace Force Development; Manpower, Personnel, and Training; Resource Management; and Strategy and Doctrine.

CONTENTS

Appendix

FIGURES

TABLES

SUMMARY

Training aircrews for combat requires access to ranges suitable for actual or simulated weapon delivery and to dedicated airspace suitable for air-to-air and air-to-ground tactics. However, a number of commercial, community, and environmental interest groups increasingly contest the access of Air Combat Command (ACC) and other military commands to ranges and airspace. To enhance this access, ACC needs a comprehensive, objective statement of its range and airspace requirements, linked to national interests, and a means to compare existing infrastructure with these requirements.

Project AIR FORCE (PAF) and ACC, working in concert, met this need by developing an analytic structure containing the following elements:

- Operational requirements that aircrews and other combatants must be trained to support.
- Training tasks required to prepare aircrews for their assigned operational tasks.
- Range and airspace characteristics needed for effective support of each training task.
- Minimum durations of training events on ranges or airspace with specified characteristics.
- Dimensions, location, equipment, operating hours, and other characteristics of current ranges and airspace.
- Relational links among operational requirements, training requirements, infrastructure requirements, and available assets.

Elements of the analytic structure are depicted in Figure S.1. Note that operational missions, objectives, and tasks are referred to collectively as a *joint mission framework.* As the figure implies, infrastructure requirements, training requirements, and the joint mission framework must be serially linked. Additionally, infrastructure requirements and current infrastructure must be linked in a way that permits ready comparisons.

This framework called for both a repository of information on various elements and a means of representing relationships among the elements. A *relational database* was the tool selected to meet these needs. The relational database constructed for this purpose contains several embedded models, developed by PAF, that automatically complete parts of the assessment process for range and airspace infrastructure. Some models transform user inputs into infrastructure requirements. Another model automatically compares

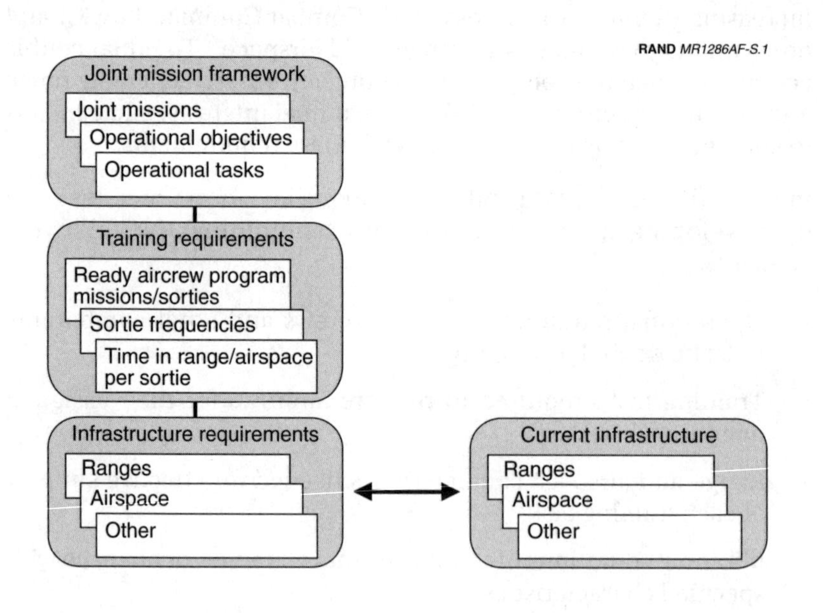

Figure S.1—The Analytic Structure

requirements with the existing infrastructure. Additionally, PAF developed several forms of graphical user interface (GUI) to facilitate user access to the database. The resulting system—a relational database combined with embedded decision models and GUI—can be referred to as a *decision support system* (DSS). The range and airspace DSS uses a Microsoft Access database and GUI of two types: Access's native capabilities and a web browser. Similarly, the models are embedded in Access relationships, queries, Visual Basic (VBA) programs, and Web server script programs.

The database contains several types of components, referred to as *objects*. These include tables, queries, forms, reports, macros, and modules. Of these, understanding the functioning of tables and queries is essential for retrieving information from the database.

Tables contain the stored information in the database. A table contains one or more *records*. Each record contains one or more *fields*. For example, we have constructed a table that lists each base with an ACC flying wing, the wing identity, and the latitude and longitude of the base. In this example, there is a record for each base. The fields are name, unit, latitude, and longitude.

Queries are used, as the name implies, to extract information from the database. They may be used to view, change, or analyze the extracted data. The most common type is a *select* query that extracts and displays selected fields from selected records in selected tables. Other types of queries may be used to build or update tables.

To realize the power and potential of the range and airspace database, a continuing investment must be made in developing and employing the human capital needed to maintain and operate it. An appropriately trained database administrator must be assigned. Staff and field users must appreciate the system's capabilities and routinely use them.

The DSS has the potential to serve a much larger staff client base than originally conceived. ACC/DOR could expand the database to include other range and airspace management information that is exchanged routinely between headquarters and field units. With some additional investment, it could be expanded to permit efficient calculations of requirements for other training resources, such as flying hours and munitions. Finally, the DSS can be expanded to include requirements and infrastructure from non-ACC range and airspace users. These might include reserve components, Air

Education and Training Command, Air Force Materiel Command, and other services.

ACKNOWLEDGMENTS

The direction and shape of this study were strongly influenced by Col Chuck Gagnon, Chief of Range, Airspace, and Airfield Management, Headquarters Air Combat Command, and Maj Michael "Buzz" Russett, our point of contact in that division at the inception of the study. Subsequent chiefs—Cols Ron Oholendt and Lynn Wheeless—provided continuing guidance and support, as did Maj Gen David MacGhee, who was Air Combat Command's Deputy Chief of Staff for Operations at a critical point in the study. Colonel Charles Hale, Lt Col Art Jean, Lt Col Frank DiGiovanni, Lt Col Dale Garrett, Maj Rob Bray, Raul Bennett, Kent Apple, and Bob Kelchner, all within the Range, Airspace, and Airfield Management Division, also made significant contributions to the project. Kent Bingham, of the Photo/Graphic Imaging Center at Hill Air Force Base, provided a graphic image of airspace for use in our cover art. During the course of our research, we visited virtually every ACC operational wing, where local range and airspace managers, operational squadron commanders, and aircrews arranged for our visits and participated in our interviews.

We are indebted to our Project AIR FORCE program directors, C. Robert Roll and Craig Moore, for their confidence and support. Leslie O'Neill, in RAND's Langley Air Force Base office, supported us generously and capably during our many visits to Headquarters Air Combat Command. Colleagues John Schank and John Stillion provided insightful reviews, and Jeanne Heller carefully edited the manuscript.

Responsibility for any remaining errors remains, of course, our own.

ACRONYMS

ABCCC	Airborne Battlefield Combat and Control Center
ACC	Air Combat Command
ACM	air combat maneuver
ACMI	Air Combat Maneuvering Instrumentation
AFI	Air Force Instruction
AGL	above ground level
AHC	advanced handling characteristics
ARTCC	air route traffic control center
ATCAA	air traffic control assigned airspace
AWACS	Airborne Warning and Control System
BFM	basic fighter maneuvers
BMC	basic mission capable
BSA	basic surface attack
CAS	close air support
CINC	commander-in-chief
C^2ISR	command, control, intelligence, surveillance, and reconnaissance

Chapter One

INTRODUCTION

BACKGROUND

Training aircrews for combat requires access to ranges suitable for actual or simulated weapons delivery and to dedicated airspace suitable for air-to-air and air-to-ground tactics. Air Combat Command (ACC) and other military commands responsible for training combat aircrews have access to an extensive inventory of ranges and airspace.

Faced with increasing competition for infrastructure usage, ACC recognized that it needed a requirements-based rather than a deficiency-based approach for determining its range and airspace infrastructure needs. In the deficiency-based approach that prevailed at the time, range and airspace resourcing alternatives were based primarily on statements of apparent *gaps* between requirements and existing capabilities. Better resourcing decisions could be made if both the requirements and current asset capabilities were stated more explicitly, with resourcing decisions based on rigorously derived assessments of the gaps.

To be defensible, infrastructure requirements must be linked firmly to training requirements, which in turn must be linked to operational requirements that demonstrably serve national interests. Additionally, for a requirements-based approach to succeed, an efficient means of comparing existing infrastructure capabilities with these vetted requirements is needed. RAND's Project AIR FORCE (PAF) was asked to help in developing these linked sets of requirements and assets.

OBJECTIVES AND APPROACH

PAF and ACC, working in concert, determined that ACC's needs could best be met through the following steps:

- Cataloging aircrew training requirements
- Relating the training requirements to operational requirements and higher-level national objectives
- Relating the training requirements to supporting range and airspace infrastructure requirements
- Comparing existing range and airspace infrastructure with requirements.

This framework called for both a repository of information on various elements and a means of representing relationships among the elements. A *relational database* was the tool selected to meet these needs. In addition to serving the analytic needs of this project, the database could be updated to reflect changes in requirements or existing assets or expanded as necessary to capture other related management information. In the hands of range and airspace managers at ACC or elsewhere, it could become a valuable tool for ongoing evaluation and management of range and airspace assets.

The relational database constructed to meet these needs contains several embedded models, developed by PAF, that automatically complete parts of the range and airspace infrastructure assessment process. Some models transform user inputs into infrastructure requirements. Another model automatically compares requirements with existing infrastructure. Additionally, PAF developed several forms of graphical user interface (GUI) to facilitate user access to the database. The resulting system—a relational database combined with embedded decision models and GUI—can be referred to as a *decision support system* (DSS).

ORGANIZATION OF THE REPORT

Chapter Two describes the elements of the analytic structure we adopted to meet ACC's needs. In Chapter Three, we provide details of the hardware and software implementation of the system. In

Chapter Four, we provide a tutorial that walks a new user through a Web-browser-facilitated tour of the database. In Chapter Five, we describe the contents of the database. Chapter Six discusses database maintenance requirements and opportunities for further development of the DSS.

Chapter Two

ELEMENTS OF THE ANALYTIC STRUCTURE

We required the following elements to fully document range and airspace infrastructure requirements, trace their relevance to training and operational requirements, and assess their adequacy:

- Operational requirements that aircrews and other combatants must be trained to support.
- Training tasks required to prepare aircrews for their assigned operational tasks.
- Range and airspace characteristics needed for effective support of each training task.
- Minimum durations of training events on ranges or airspace with specified characteristics.
- Dimensions, location, equipment, operating hours, and other characteristics of current ranges and airspace.
- Relational links among operational requirements, training requirements, infrastructure requirements, and available assets.

These elements relate to each other in an analytic structure that is depicted in Figure 2.1. Operational missions, objectives, and tasks are referred to collectively as a *joint mission framework*. As the figure implies, infrastructure requirements, training requirements, and the joint mission framework must be serially linked. Additionally, infrastructure requirements and current infrastructure must be linked in a way that permits ready comparisons.

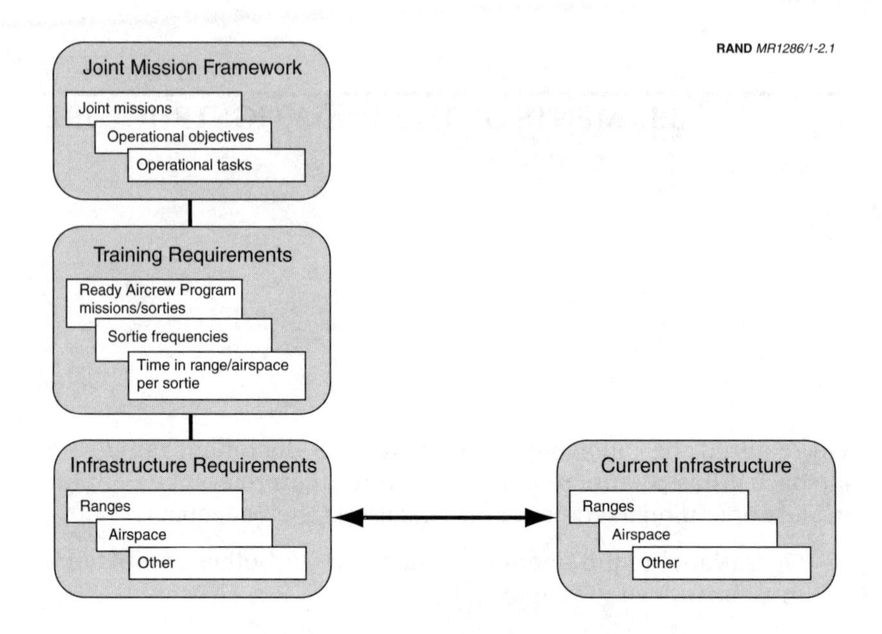

Figure 2.1—The Analytic Structure

We next describe the elements of the range and airspace analytic structure and document how we developed and populated the various elements and linkages.

OPERATIONAL REQUIREMENTS: THE JOINT MISSION FRAMEWORK

In developing this framework, we sought to express commanders-in-chief's (CINCs') warfighting needs in terms of desired operational effects. Using a strategies-to-tasks concept, we developed a set of operational missions, objectives, and tasks to describe how military power can be applied jointly. The framework contains 11 joint operational missions that collectively describe the broad outcomes CINCs seek to achieve in operations ranging from major theater war to smaller-scale peacekeeping and peacemaking contingencies. Within these missions, we identify some 40 operational objectives and 150 operational tasks.

TRAINING REQUIREMENTS: AN ADAPTATION OF THE READY AIRCREW PROGRAM

The next element in the analytic structure is a representation of training activities needed to prepare aircrews to support operational requirements. To complete the linkages envisioned in the analytic structure, training activities must be related, on one hand, to operational requirements, and on the other hand, to training resource needs, specifically range and airspace infrastructure.[1]

The DSS focuses primarily on mission qualification and continuation training. Undergraduate flying training and initial qualification training are accomplished primarily through formal training courses and generally do not place demands on ACC range and airspace infrastructure.[2] Special mission and upgrade training is often accomplished using sorties that are dual-logged as continuation training. Thus, demand for ACC ranges and airspace is largely a function of mission qualification and continuation training requirements.

Mission qualification and continuation training requirements are outlined in the Air Force's Ready Aircrew Program. RAP requirements are contained in mission design series (MDS)-specific, 11-2 series Air Force Instructions (AFIs) and in annual tasking messages published by ACC. For aircrews in each MDS, RAP specifies a total number of sorties per training cycle, broken down into mission types, plus specific weapon qualifications and associated events. The specified number of sorties varies depending on the aircrew member's experience and qualification level.[3] For example, mission

---

[1]Readers should not infer that training requirements used in our analysis were derived from our joint mission framework. We derived our training requirements from the Air Force's Ready Aircrew Program (RAP), as described below, which in turn is derived from other representations of operational requirements such as unit designed operational capability (DOC) statements. We then linked our training requirements framework to our joint mission framework.

[2]For a few weapon systems, ACC does conduct initial qualification training. These programs usually are co-located with at least one combat squadron and must share local training infrastructure. This study did not include the initial training requirements for these systems; therefore, the total requirement for these bases is underestimated.

[3]*Experienced* pilots have accumulated a specified number of flying hours. For example, fighter pilots are considered experienced if they have accumulated 500 hours

category sorties for one type of F-16 for the 1998–1999 RAP cycle are shown in Table 2.1.

Table 2.1

RAP Mission Category Sorties for the F-16CG

	Annual Sortie Requirement			
	Basic Mission Capable		Combat Mission Ready	
Mission Category	Inexperienced	Experienced	Inexperienced	Experienced
Basic surface attack				
(BSA)(day)	6	4	8	6
(BSA (night)			4	3
Surface attack tactics				
(SAT) (day)	6	4	14	12
(SAT (night)			4	3
Close air support (CAS)			4	3
Defensive counter air				
(DCA) (day)	3	2	10	8
(DCA (night)			4	2
Air combat maneuver (ACM)			8	6
Basic fighter maneuver (BFM)	3	2	8	6
Red air (opposing force for air-opposed training sorties)			8	8
Commander option	54	48	18	19
Total	72	60	90	76

in their primary aircraft, or 1000 total hours of which 300 are in their unit's primary aircraft, or 600 fighter hours of which 200 hours are in their unit's primary aircraft, or who reached an experienced level in another fighter MDS and have 100 hours in their unit's primary aircraft.

Line pilots in operational units generally attain a qualification level designated *combat mission ready* (CMR). Pilots in staff positions generally attain a lower level of qualification designated *basic mission capable* (BMC).

In each training cycle, RAP specifies more sorties for inexperienced aircrew members than for experienced aircrew members and more sorties for CMR qualification than for BMC qualification. Additionally, RAP may specify more sorties for active component aircrews than for reserve component aircrews.

Sortie Types Used in the Analysis

RAP sorties may be either *basic* or *applied*. Basic sorties are building-block exercises, such as air handling characteristics (AHC), basic surface attack, or basic fighter maneuver, that are used to train fundamental flying and operational skills. Applied sorties, such as surface attack tactics and defensive counter air, are intended to more realistically simulate combat operations, incorporating intelligence scenarios and threat reaction events.

The examples of sortie types in the preceding paragraph are all fighter-oriented. For nonfighter aircraft, basic sorties are generally identified as combat skills sorties (CSS). Applied sorties for nonfighter aircraft are generally identified as SAT sorties (bombers) or *mission* sorties (for aircraft that do not deliver weapons).

For our analysis, we generally used the RAP sortie structure and annual sortie requirements as a statement of training requirements. However, in some cases, notably SAT, we subdivided RAP sorties into several types (which we refer to as *variants*) that differ significantly from each other in their infrastructure requirements. For example, fighter SAT missions are divided into *air opposed, ground opposed,* and *live ordnance* variants. Similarly, SAT missions for bombers are divided into *inert high/medium level, inert low level, live ordnance, simulated delivery of ordnance,* and *maritime* variants.

We also postulated the need for sorties that combine several MDS, performing different operational roles, in a single training mission. For squadron-size exercises, we used the term *large force engagement* (LFE) to identify these sorties. For less than squadron-size exercises, we used the term *small multi-MDS engagement* (SMME). We refer to LFEs and SMMEs collectively as *combined* sorties. We did not develop a comprehensive list of SMMEs. However, we have structured several examples that suggest the possibilities of specifying such training requirements. In general, RAP does not currently specify multi-MDS sorties except for LFE requirements in some MDS and a few exceptions such as a forward air control aircraft (FAC-A) working in conjunction with CAS aircraft.

A complete list of the sortie types used in our analysis and their categories and definitions can be found in Appendix A.

Relating Training Requirements to Operational Requirements

We determined that applied and combined sorties would be related directly to operational tasks found in our joint mission framework. Basic sorties and variants would be related to various applied sorties, and could then be related indirectly to operational tasks. The relationships are shown in Figure 2.2. Matrices relating various MDS/sortie combinations to specific operational tasks within the joint mission framework are too large to be readily included here. However, the linkages are reflected in the range and airspace database we constructed and can be extracted for any MDS, sortie type, or joint mission. As an example, the joint mission "Deny the enemy the ability to operate ground forces" contains an operational objective "Halt invading armies," within which one of the operational tasks is "Delay/destroy/disrupt lead units of invading armies." For the F-16CG, the database associates this operational task with three types of applied sorties (CAS, DCA, and LFE) and five types of basic sorties (instrument [INS], AHC, ACM, BFM, and BSA). Range and airspace infrastructure required for the F-16CG for each of these sortie types can also be extracted from the database and linked to this operational task.

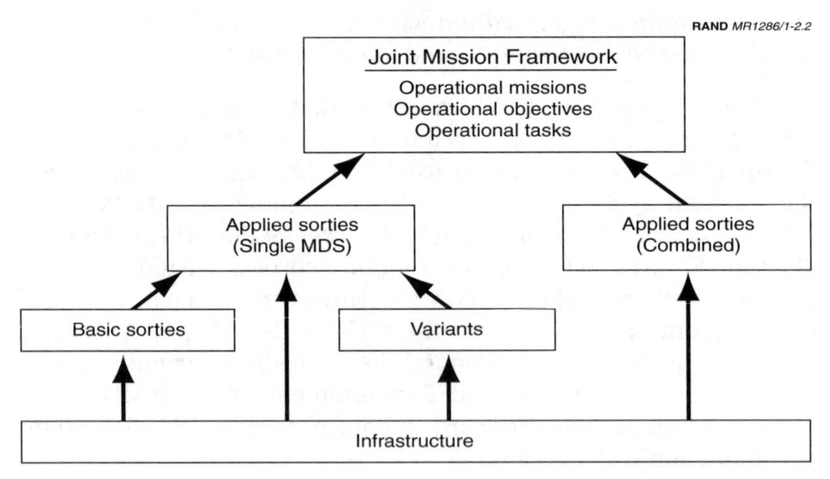

Figure 2.2—How Sorties Link Infrastructure to Operational Tasks

REQUIRED INFRASTRUCTURE CHARACTERISTICS

The next element in our analytic system is a statement of the range and airspace infrastructure needed to support training requirements. To be useful for training, the range and airspace infrastructure must have certain geographical, qualitative, and quantitative characteristics. Geographically, it must be reasonably proximate to base operating locations. For many MDS, especially fighters, extending aircraft range through air refueling is not a viable option for training sorties. Even for longer-legged bomber and command, control, intelligence, surveillance, and reconnaissance (C^2ISR) aircraft, other constraints such as crew duty-day length and flight time engaged in useful training versus time spent cruising to and between training areas need to be considered. Qualitatively, the infrastructure must have minimum dimensions, equipment, authorization for operating aircraft and systems in specified ways, and other characteristics. Quantitatively, the time available on proximate ranges and airspace must be sufficient to support the training requirements at an operating base. In this and the following sections, we discuss how these infrastructure requirements were developed and are represented in the range and airspace information system.

Distance from Base to Range/Airspace

Ranges and airspace must be reachable with the maximum fuel load consistent with the sortie type. Further, fuel available for cruising to, from, and between ranges and airspace must take into account the amount of fuel consumed during training events.[4] Because many sortie types require access to more than one asset (e.g., a low-level route, a maneuver area, and a range) during a given sortie, the required geographical proximity of the assets cannot be adequately expressed in terms of a radius from the base. It is better expressed in terms of a maximum for the sum of the free cruising legs between

---

[4]We use the term *training event* to indicate a part of a sortie with a specific training focus. For example, an air-to-ground sortie may include a low-level navigation leg, a threat evasion exercise, and a series of weapon deliveries. In our usage, these training-related components of the sortie are referred to as a training event.

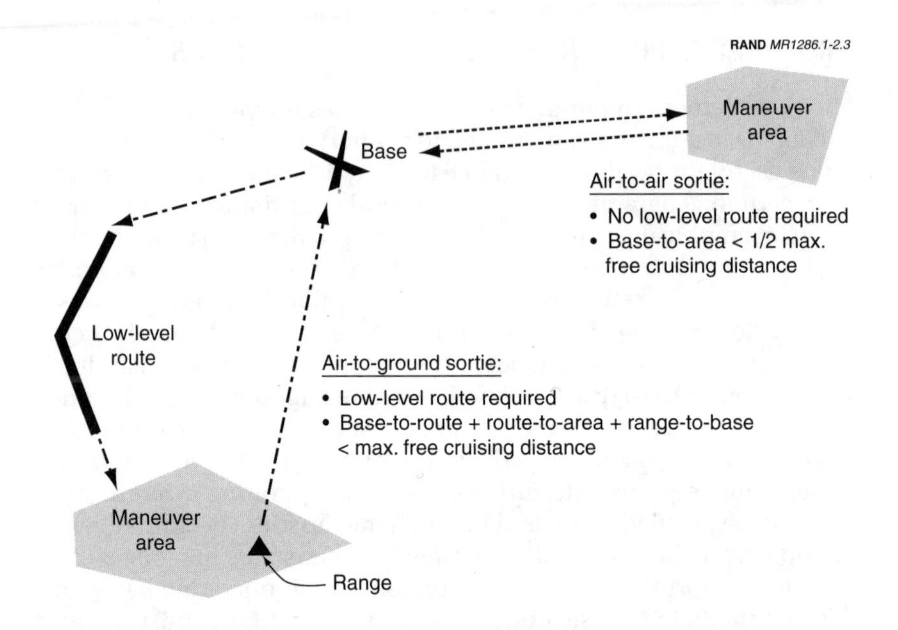

Figure 2.3—Maximum Distance from Base to Range/Airspace

assets (see Figure 2.3). We calculated this maximum for each MDS/sortie-type combination and used it to analyze the geographical relationships of bases, ranges, and airspace.

To calculate the maximum free cruising distance for each MDS/sortie-type combination, we interviewed aircrew members to determine the normal external fuel-tank configuration and fuel capacity for the sortie; fuel consumption for taxi, takeoff, and climb; fuel consumption during training events en route, in the area and/or on the range; and reserve fuel requirements. To determine fuel consumption, we first determined standard minimum durations for each training event. We determined reasonable values for these minimums through consultation with experienced aircrew members.

Subtracting required consumptions and reserve from fuel capacity yields the amount of fuel that can be used for free cruising legs. Dividing this amount by an average fuel consumption rate at a typical cruising speed and altitude yields the maximum free cruising time. Multiplying this time by the typical cruising speed gives the maximum free cruising distance. Maximum free cruising distances ranged from 79 miles for F-15C BFM sorties to 1757 miles for B-52 SAT sorties. In general, fighters and helicopters are far more limited in their free cruising distances than are bombers and C^2ISR platforms.

Qualitative Requirements

Qualitative infrastructure requirements (e.g., range dimensions, equipment, operating authorizations) were developed primarily through capturing the judgment of experienced aircrew members. In general, we used the MDS/sortie-type combination as the unit of analysis (i.e., each MDS/sortie-type combination would have its own unique set of infrastructure requirements represented in the database). However, where choice of events would significantly alter the infrastructure requirement, we divided the RAP sortie into two or more variants, as discussed in the previous section of this chapter. This enabled us to better reflect specific infrastructure standards for the wide variety of crew activity being logged under any one sortie type.

As an exception to the general rule of using the MDS/sortie-type combination as the unit of analysis, we found that for range characteristics related to weapon deliveries, it was necessary to use the training event (i.e., the weapon delivery type) as the unit of analysis. Weapon deliveries can vary by release altitude, release type (level, loft, dive, etc.), weapon type (rocket, gravity bomb, guided munition, etc.), level of threat (which affects assumed accuracy of the delivery), and MDS.[5] Weapon delivery type affects two categories of range characteristic requirements—restricted airspace dimensions and weapon safety footprint area (WSFA) dimensions. To specify standard range requirements for weapon deliveries at an MDS/ sortie-type level of analysis, we would have to identify the most demanding (in terms of these range characteristics) weapon deliveries that aircrews should routinely employ in each MDS/sortie combination. However, we found no basis for selecting which weapon delivery types should be used to set these requirements.

[5]ACC currently identifies 210 distinct weapon delivery types.

Thus, restricted airspace and WSFA requirements are expressed at the event rather than the MDS/sortie-type level in our analysis.

Organizing the Qualitative Requirements

Qualitative requirements (and corresponding information on existing assets) were captured for six infrastructure types: low-level routes, maneuver areas, ranges, threats, orbits, and other. Specific characteristics appearing in these requirement arrays are listed in Appendix B.

This organization was developed to state the need for infrastructure without being limited to current airspace terms such as restricted area, military operation area (MOA), warning area, air traffic control assigned airspace (ATCAA), or military training route (MTR). These terms are for the most part derived from the air traffic control lexicon rather than a training lexicon. Moreover, training requirements can often be met by any of several current airspace types, or, as is frequently observed, they may require combinations of several airspace types. Thus, we sought to define the infrastructure requirements using generic terms, e.g., low-level route rather than MTR, maneuver area rather than MOA.

Low-Level Routes. Air-to-ground sorties are generally required by training publications (AFI 11-2 series) to incorporate a low-level ingress route. An MTR typically connects to a MOA surrounding a range. The length of the route, its required altitudes, and other required attributes are captured in the range and airspace database.

Maneuver Areas. Air-to-ground sorties may require controlled airspace for attack tactics and threat reaction, generally requiring a MOA and perhaps a vertically adjacent ATCAA. Air-to-air sorties also require a maneuver area—either a MOA with an ATCAA or an offshore warning area. Required vertical and lateral dimensions and other attributes of the maneuver area are captured in the range and airspace database.

Ranges. A range is required for air-to-ground sorties. Ranges also require restricted airspace over their targets large enough to contain released weapons and the long and cross dimensions of weapon safety footprints. Required vertical and lateral dimensions of the restricted area, types of targets, scoring systems, and other related range attributes are specified in the range and airspace database. The relationship of weapon safety footprints, WSFAs, and restricted airspace is illustrated in Figure 2.4.[6]

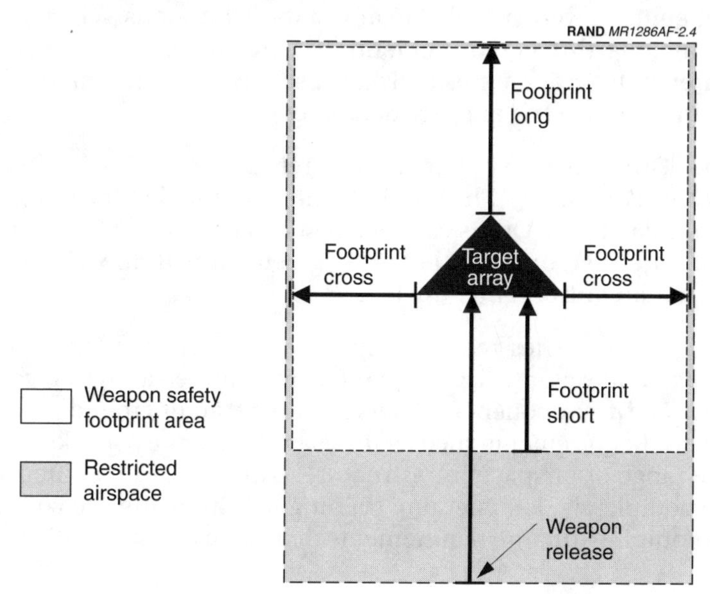

Figure 2.4—Weapon Safety Footprints, Weapon Safety Footprint Areas, and Restricted Airspace

[6]To calculate WSFA and restricted airspace requirements, RAND used (1) weapon safety footprint data for 210 distinct delivery types, obtained from ACC/DOR in August 1999, (2) an assumed target array size of 2 nm × 2 nm, and (3) weapon release points calculated using Combat Weapons Delivery Software (CWDS) provided by the Mission Planning Support Facility, OO-ALC/LIRM, Hill Air Force Base, UT.

Note that Figure 2.4 provides the WSFA and restricted airspace requirements for only a single axis of attack. For multiple axes of attack, the dimensions shown in Figure 2.4 must be rotated around the target.

Threats. Many air-to-ground sorties require ground-based radar threat emitters or communications jammers, which may be installed on a range, beneath a MOA, or conceivably at points along an MTR. We determined that the training requirement would be met if the threat emitters were installed in any of these locations. Thus, rather than include threat requirements within range, area, and route requirements arrays, we established a separate threat requirements array in the range and airspace database.

Orbits. Orbits may be required for air refueling or certain command and control missions. The requirement is captured in the range and airspace database. Orbits can be flown in a MOA or ATCAA, but are usually specified only in a letter of agreement with the affected air route traffic control center (ARTCC).

Other. Some sorties require a specific other aircraft for effective training. For example, DCA and offensive counter-air (OCA) sorties require *Red air* opponents. Others require an air or ground weapons director. Requirements such as these are not, strictly speaking, part of the range or airspace infrastructure. However, in the interest of more completely documenting training requirements, we collected such noninfrastructure requirements that came to our attention.[7]

Capacity

The amount of operating time required on ranges and in airspace can be calculated, for a given MDS/sortie-type combination, by multiplying the required number of sorties by the time required for an individual sortie on a range and/or in an airspace. After certain adjustments (discussed below), the results can be summed across all MDS/sortie-type combinations to determine a base's total local demand for ranges and airspace (referred to as assets in Figure 2.5 and Chapter Six). This demand is computed and recorded in the range and airspace database for each base/MDS/sortie-type

[7]Pilots we interviewed said that training with other MDS is very important, but the lack of a *requirement* for such training often discouraged an already-busy potential "partner MDS" from participating.

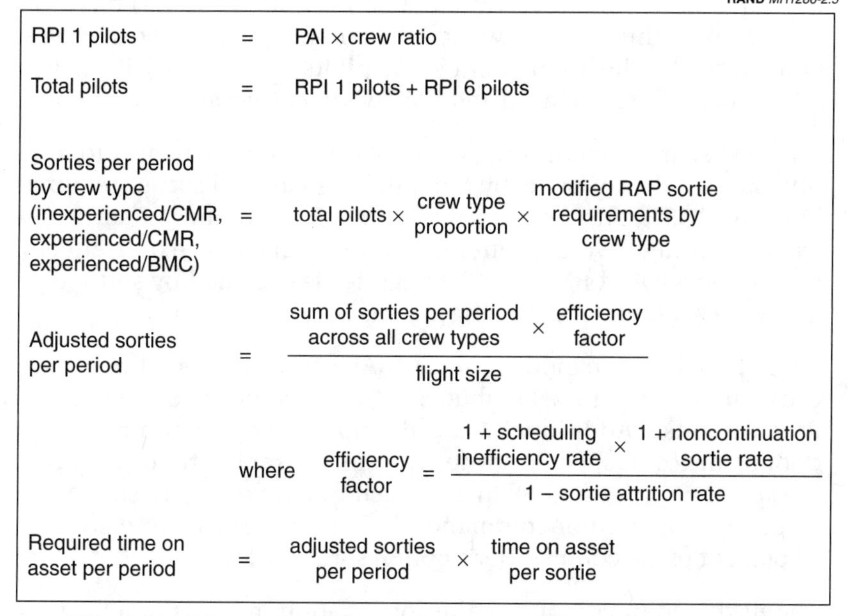

Figure 2.5—Determining Sortie and Time on Asset Requirements

combination. In the following paragraphs, we discuss, first, how the required number of sorties is calculated and, second, how the time required for each sortie is determined.

Required Number of Sorties. The database contains a table that lists the total number of annual sortie requirements by base, MDS, and sortie type. To populate this table, we determine the number of pilots in each MDS at each base and multiply that number by the annual requirement for each sortie type.[8] The required calculations are shown in Figure 2.5 and described below.

To determine the number of pilots, we first obtain the primary mission authorized inventory (PMAI) by MDS and base. These counts are multiplied by the crew ratio for the MDS, yielding the

[8]In some MDS, crew positions other than pilot also require training. However, we found no MDS with a crew position that required more sorties than did the pilot. Thus, using pilot counts alone (excluding co-pilots) as the basis for annual sortie requirements is sufficient to establish an upper bound on sortie demand.

expected number of RPI 1 (RPI = rated position identifier) pilots on the base.[9] To this number, we add the number of RPI 6 pilots by base and MDS.[10] The total number of pilots is then distributed to experienced/inexperienced and BMC/CMR categories.[11]

The next step in determining the total sortie requirement is to multiply the number of pilots by the number of annual sorties required in each MDS/sortie-type combination. The number of sorties in each training cycle (generally one year) for experienced/inexperienced and BMC/CMR categories is specified by sortie type and MDS in annual RAP tasking messages.

For our analysis, we modify the raw RAP counts in several ways. We use assumed rates to redistribute RAP sortie counts to our modified-RAP variants. Additionally, we distribute the the commander's option sorties to specific sortie types in the same proportions that the specific sorties had relative to each other, i.e., if SAT sorties are 40 percent of the noncommander's option sorties, we distribute 40 percent of the commander's option sorties to SAT.[12]

The next step in computing the sortie requirement is to adjust for flight size. When two-ship or four-ship flights use a range or airspace, multiple aircrews obtain training in the same time period. Thus, the critical factor in quantifying range and airspace demand is not the annual number of sorties but rather the annual number of flights. To convert sortie counts to flight counts, we divide sortie counts by an assumed average flight size for each MDS/sortie-type combination.

---

[9]RPI 1 identifies line pilots (excluding commander and operations officer) occupying cockpits in operational squadrons.

[10]RPI 6 identifies commanders, operations (ops) officers, and pilots in staff positions.

[11]For these calculations, we consider RPI 6 positions, except commander and ops officer, to be experienced and BMC. Commander and ops officer are considered experienced and CMR. RPI 1 pilots are considered CMR and are distributed using assumed rates between experienced and inexperienced categories.

[12]RAP specifies the number of sorties by type that each aircrew member must fly in a training cycle. Additionally, it specifies a number of sorties that can be of any type, depending on the commander's judgment of where the individual or unit needs training emphasis.

The final step in developing and adjusting the sortie requirement is to inflate the count to account for attrition (maintenance and weather cancellations), scheduling inefficiency, and noncontinuation training sorties. Some scheduled sorties cannot be completed because of either maintenance or weather aborts. Although these aborted sorties do not satisfy training requirements, they nonetheless consume available time on ranges and airspace because the scheduled time generally cannot be reallocated on short notice (in the case of maintenance aborts) or used by other aircrews (in the case of weather or mission conflict aborts).[13] A scheduling inefficiency factor accounts for the fact that perfectly efficient scheduling, using 100 percent of available range or airspace time, would tend to suboptimize overall aircrew time management because it would adversely affect aircrew workday and work/life balance considerations. Finally, some but not all upgrade and special qualification sorties are dual-logged as RAP sorties. The noncontinuation training inflation factor builds a range/airspace infrastructure requirement for upgrade and special qualification sorties that are not dual-logged. The range and airspace database uses assumed values for these three factors (10 percent for each factor).

Time Required per Sortie on Range and/or in Airspace. A table indicating time required per sortie on a range or in an airspace, by MDS and sortie type, is found in the database. See, for example, times in Table 2.2. The times shown in this table (minimum training event durations) are assumed values based on interviews with Weapons School and operational unit aircrews. They represent minimums considered necessary for the sortie to produce some standardized training value.

Total Demand. Total range and airspace time requirements by base, MDS, and sortie type are calculated and reflected in a table in the database. Table 2.2 reflects, for example, an extract of this part of the

[13]A few units fly a large number of sorties on ranges that they do not control, which can result in a mission conflict. Usually, once a unit contracts for time on a range, there is little chance of mission conflicts with the owning unit. However, we found at least one range (White Sands Missile Range Complex) where the range time could be canceled by range controllers within 15 minutes before entry time. In this case, fighter aircraft are already airborne when they are canceled.

Table 2.2

Infrastructure Demand: F-16CGs at Hill AFB

Sortie Type	Total Annual Sorties	Time per Sortie (minutes)	Average Flight Size	Annual Required Infrastructure Time (hours)
BFM	752	40	1	674
BSA	1,128	40	2	506
CAS	357	50	2	200
DCA	1,203	35	4	236
SAT	1,474	35	4	289
SEAD-C[a]	184	30	4	31

[a]Suppression of enemy air defense.

database for F-16CGs at Hill AFB. This requirement can be interpreted as a demand for maneuver airspace time for air-to-air sortie types and as a demand for both maneuver airspace and range time for air-to-ground sortie types. It is determined, as shown in Figure 2.5, as the product of total requirements for a given base/ MDS/sortie-type combination multiplied by the time required on asset for that MDS/sortie-type combination.

Data Limitations. Lack of available empirical data and other related problems required us to estimate many of the factors used to compute capacity requirements. A discussion of these limitations is provided in Appendix C.

CURRENT INFRASTRUCTURE

Information regarding the characteristics of ranges and airspace commonly used by ACC aircrews was collected (by e-mail) by ACC/DOR during late 1998 and early 1999. Preformatted Excel spreadsheets were sent as attached documents to local range managers and airspace schedulers, who entered the required information in the spreadsheets and returned them to ACC/DOR. The spreadsheets were subsequently forwarded to PAF to be incorporated in the database. Subsequently, a capability was provided to permit local range managers and airspace schedulers to update these characteristics via a Web interface. Specific characteristics tracked in the range and airspace database are listed in Appendix B. They can be found in various tables in the database and in selected displays available via a Web browser. Limitations on the available data are discussed in Appendix C.

COMPARISON OF CURRENT INFRASTRUCTURE WITH REQUIREMENTS

An important element of our analytic structure is a capability to compare requirements and resources. Linkages and models embedded in the range and airspace database permit current infrastructure and requirements to be compared for each MDS/sortie-type combination. These comparisons are reflected in a series of tables in the database and in a display accessible via a web browser. The example from the web browser shown in Chapter Five, Figure 5.3, depicts an assessment of maneuver areas for F-15C DCA sorties. Each row represents a different maneuver area (identified in the "name" column). Characteristics of the various areas are shown under "width," "length," etc. Characteristics that meet requirements are shaded light gray (green on the web) while those that do not meet requirements are shaded dark gray (red on the web). This screen depicts only part of a much larger matrix containing all areas and all characteristics of areas.

Chapter Three

ELEMENTS OF THE DECISION SUPPORT SYSTEM

A CONCEPTUAL DESCRIPTION OF THE DSS

A DSS provides measures of performance that a decisionmaker can use with his own expertise in making a decision. The measures of performance require one or more models that capture intrinsic relationships. In the present case, the models capture the relationships among operational requirements (a joint mission framework), aircrew training requirements, range and airspace infrastructure requirements, and available infrastructure.

To do their job, the models must include data on the joint mission framework (JMF), on which training sorties contribute to the JMF objectives and tasks, and infrastructure requirements and assets. These data are captured in formal databases. On the other hand, the interactions between the DSS and the user need to be facilitated by a user-friendly interface: a graphical user interface (GUI) similar to that used in Macintosh and Windows software. Overall, a DSS requires databases, models, and GUI.

The database for the range and airspace DSS is implemented in Microsoft Access. The models included in the DSS are embedded in Access relationships, queries, Visual Basic (VBA) programs, and web server script programs. The DSS uses GUI of two types: Access's native capabilities and a web browser. These elements of the DSS are discussed in more detail below.

DATABASE AND MODELS EMBEDDED IN MICROSOFT ACCESS

Data tables and relationships among the tables are the most prominent features of Access, but Access also incorporates features such as queries, semantic information (e.g., description fields), and user-defined functions and subroutines (in VBA modules). In the range and airspace DSS, these relationships, queries, functions, and subroutines are used to create models that help estimate infrastructure requirements and then help compare the required with the actual infrastructure by creating a color-coded chart that highlights which available infrastructure can or cannot meet specific requirements. The web interface to the DSS also obtains its data from the Access database, as noted in the web section below.

USER INTERFACE VIA ACCESS

Access's GUI facilitates updating the various tables in the database through views similar to grids or spreadsheets. This GUI also can display a diagram of table relationships (via its Tools -> Relationships menu).

Furthermore, Access provides a diagrammatic display of relationships embedded within each query, complementing the relationships diagram noted above. For the analyst with a good understanding of Access and the range and airspace DSS, Access simplifies the creation of ad hoc queries that may provide new measures of performance useful in assessing the current infrastructure.

Finally, Access's GUI simplifies the documentation of database, tables, and models by making available its various description fields.

USER INTERFACE VIA A WEB BROWSER

Access is sufficient to share data among a few users at one location, but it becomes more cumbersome when retrieval and update of data from multiple locations are required. Local base range and airspace managers are integral parts of the infrastructure system being modeled. To facilitate their use of the DSS, tools such as web browsers and a server are required.

A web server is a program that runs at a central location and allows the viewing of data and graphics by a large number of users in different locations. The web server is the program that responds to requests from each user's web browser to display or update specified data. In the range and airspace DSS, the web server pulls data directly from the Access database.

The query capabilities provided by the web browser/server interface are quite limited relative to the full ad hoc query capabilities of Access. Thus, some users will want to obtain the full database for query purposes rather than relying on web-based queries.

HARDWARE AND SOFTWARE REQUIREMENTS

The system was designed to operate with hardware and software generally available in Air Force desktop systems and server environments. Detailed specifications are provided in Appendix D.

Chapter Four

WHAT IS IN THE DATABASE

A database may contain several types of components, referred to as *objects*. The types of objects found within Access are tables, queries, forms, reports, macros, and modules. Of these, an understanding of tables and queries is essential for retrieving information from the database.

In this chapter, we provide an orientation to the major tables and queries and a brief discussion of the use of forms, reports, macros, and modules in the database. Our purpose here is to assist a user who is familiar with Access to rapidly gain an ability to *extract information* from the database. This orientation is not designed to prepare a user to *maintain* the database. Additional training may be needed to prepare a user for maintenance responsibilities.

TABLES

Tables contain the stored information in the database. A table contains one or more *records*. Each record contains one or more *fields*. For example, we have constructed a table that lists each base with an ACC flying wing, the wing identity, and the latitude and longitude of the base. See Figure 4.1. In this example, there is a record for each base. The fields are name, unit, latitude, and longitude.

RAND *MR1286.1.4.1*

Base	Unit	Latitude	Longitude
BARKSDALE	2 BW	32.5003	-93.6633
BEALE	9 WG	39.1361	-121.4367
CANNON	27 FW	34.3833	-103.3217
DAVIS-MON	355 WG	32.1650	-110.8817
DYESS	7 BW	32.4200	-99.8567
EGLIN	33 FW	30.4867	-86.5267
ELLSWORTH	28 BW	44.1450	-103.1033
HILL	388 FW	41.1267	-111.9700
HOLLOMAN	49 FW	32.8517	-106.1017
KEFLAVIK		64.9853	-22.6056
LANGLEY	1 FW	37.0833	-76.3617
MINOT	5 BW	48.4150	-101.3567
MOODY	347 FW	30.9683	-83.2767
MTN_HOME	366 WG	43.0433	-115.8700
NELLIS	57 WG	36.2367	-115.0333
OFFUTT	55 WG	41.1183	-95.9125
POPE	23 WG	35.1700	-79.0150
ROBINS	19 ARW	32.6403	-83.5919
SEYMOUR_J	4 FW	35.3400	-77.9600
SHAW	20 FW	33.9733	-80.4733
TINKER	552 ACW	35.4183	-97.3892
WHITEMAN	509 BW	38.7300	-93.5483

Figure 4.1—Sample Table from the Database

Master Lists

One group of tables provides master lists of key variables in the database. One such table, labeled *tblMDS,* lists all mission design series (MDS) for which training and infrastructure requirements have been identified and provides a description and other information regarding the MDS. Another such table, labeled *tblSortie,* lists all mission/sortie types and defines each. In addition to providing a repository for definitions and other basic information related to the key variables, these tables also ensure consistent terminology throughout the database. In other tables, whenever an MDS or sortie field is used, the database ensures that each entry in the MDS or sortie field matches an entry in the MDS or sortie master lists. This is called *referential integrity.* It ensures that the same item is not inadvertently identified in two different ways within the database (e.g., "F15E" vs. "F-15E").

The Joint Mission Framework

Another set of tables lists the joint missions, operational objectives, and operational tasks that constitute the joint mission framework. The tables used for this purpose are *tblJntMsn, tblOpOb,* and *tblOpTsk.* In addition, there is a table named *tblOpOb_OpTsk,* which we refer to as an *intersection* table. The existence of a record in *tblOpOb_OpTsk* for a particular operational objective and operational task indicates a relationship between the two: the task supports the objective. Operational objectives and operational tasks have a *many-to-many* relationship, meaning that an operational objective may have more than one operational task associated with it and an operational task may have more than one operational objective associated with it. In fact, almost all operational objectives have more than one operational task associated with them, but only a very few operational tasks are associated with more than one operational objective.

In contrast, joint missions and operational objectives have a one-to-many relationship. Each joint mission has one or more operational objectives associated with it, but each operational objective is associated with one and only one joint mission. Because of their one-to-many relationship, an intersection table was not required to relate joint missions to operational objectives. This relationship is shown by providing a joint mission field in *tblOpOb.*

Training Requirements

To relate training requirements to the joint mission framework, it is necessary to link the mission/sorties listed in *tblSortie* to the operational tasks listed in *tblOpTsk.* For *applied sorties,* the linkage to operational tasks is accomplished directly, using the intersection table *tblOpTsk_AppSor.* Other sortie types are linked indirectly, through their relationship to applied sorties. *Basic* sorties are related to applied sorties in *tblAppSor_BasSor,* and combined sorties are related to applied sorties in *tblMMEspecs.* Although not used in the current analysis, a framework for analyzing a training sortie in terms of its included events can be found in *tblTrnEvt*.

Quantifying Demand

Data used to quantify the demand for infrastructure are spread among a number of tables, as indicated in Table 4.1.

Infrastructure Requirements

To determine the infrastructure requirements for a given MDS/sortie combination, one refers to the appropriate record in *tblMDS_Sortie_Infra*. This record refers, in turn, to a record in *tblInfraStruct* that

Table 4.1

Information Required to Quantify Infrastructure Demand

Information	Table
Number of PMAI aircraft by MDS and base	*tblPMAI*
RAP sortie requirements by pilot class	*tblRapSortieReqmts*[a]
Weights used to distribute RAP sorties to the modified-RAP sortie definitions used in the database	*tblRapSortieDistribution*
Modified-RAP sortie requirements by pilot class[b]	*tblRaSortieReqmts*
Crew ratios	*tblCrewRatio*
RPI 6 inventories	*tblRpi6Inventory*
Total modified-RAP sortie requirements by base and MDS	*tblMdsBaseRaSortie*
Duration of training time on range or in airspace, average flight size, sortie attrition rate, scheduling inefficiency rate, and noncontinuation training sortie rate, by MDS and modified-RAP sortie	*tblSor_Mds_Infra*

[a]This table was developed with infrastructure resource demand in mind, and as such should be used with caution for other purposes. For example, for F-15C, there is a significant Red Air sortie allocation in the RAP tasking message that is not reflected in *tblRapSortieReqmts*. If this were included there would be a double-counting of infrastructure demand—once for the F-15Cs and once for the aircraft they oppose as Red Air.

[b]See discussion of modified-RAP sortie types under the heading "Sortie Types Used in the Analysis" in Chapter Two. Note that we use the segment "Rap" in the labels of tables and queries that contain original RAP sortie information and the segment "Ra" in labels of tables and queries that contain modified-RAP sortie information.

acts as a directory record to tables containing requirements for various kinds of infrastructure.[1] These templates appear as records in the series of tables labeled *tblTpRoute, tblTpArea, tblTpRange, tbl TpOrbit, tblTpThrtEC,* and *tblTpOther* ("Tp" indicates "template").[2] Additionally, *tblFootprints* provides dimensional requirements for each type of weapon delivery event.

Current Infrastructure

Information on current infrastructure is contained in a series of tables labeled *tblInfraItemRoute, tblInfraItemArea, tblInfraItem Range,* and *tblInfraItemOrbit.*

Infrastructure Evaluations

A series of tables contains the results of comparing each current infrastructure item with the requirements for each MDS/sortie combination. Entries in these tables indicate whether or not the infrastructure meets the requirement. These tables are labeled *tbl MatchingRoute, tblMatchingArea, tblMatchingRange,* and *tbl MatchingOrbit.* Additionally, *tblMatchingRangeDelivery* provides comparisons of each weapon delivery event's requirements with each range's characteristics.

QUERIES

Queries are used, as the name implies, to extract information from the database. They may be used to view, change, or analyze the extracted data. The most common type is a *select* query that extracts and displays selected fields from selected records in selected tables. Other types of queries may be used to build or update tables.

---

[1] *tblMDS_Sortie_Infra* does not serve this directory role because a number of MDS/sortie combinations could have a common infrastructure requirement. When this is the case, they all refer to the same infrastructure record in *tblInfraStruct.*

[2] For similar multiple-use reasons, the individual infrastructure templates do not appear directly within the rows of *tblInfraStruct.* Perhaps more important, the hierarchical layering serves to organize the infrastructure requirements and facilitates comparison with actual training infrastructure resources.

Several standard queries have been constructed for recurring database maintenance tasks or for anticipated recurring analysis needs. Some of these are listed in Table 4.2.

Table 4.2

Selected Standard Queries

Query Name	Information or Process
tbl MDS_Infra	List of all infrastructure requirements, by MDS and modified-RAP sortie
tbl OpObj_MDS	List of operational objectives and tasks associated with a selected MDS or sortie type
tbl Op Tsk_Basic Sortie	A complex inquiry that relates basic sorties and variants indirectly, through applied sorties, to operational tasks
qselComputeInfraTimeReqmt	Computation of time required on range or asset for each base/MDS/sortie combination
quniMdsBase	Gives pilot experience/qualification breakout for each MDS/base combination. A complex query that pulls information from multiple sources. Experienced BMC pilots come from the RPI 6 data in *tblRpi6Inventory*. CMR pilots are computed from the crew ratio and experience levels in *tblMds*, and the PMAI values in *tblPMAI*. This query result has the logical status of a table.
qmakMdsBaseRaSortie	Combines the pilot class and pilot count information in *quniMdsBase* with the sortie requirements by pilot class in *tblRaSortieReqmts* to construct the sortie requirement for the base. Broken out by MDS and sortie type for each base.
qselRaSortieReqmts	Distributes the RAP sortie requirements from *tblRapSortieReqmts* onto the modified-RAP sorties using the weights in *tblRapSortie Distribution*. Grouped by MDS, sortie, and pilot class.
qmakBaseRange	This combines *tblBase* with *tblInfraItemRange* to create *tblBaseRange*, which includes the distance between the base and range. It uses the macro *SeparationLatLong*.
qselRaSortieReqmts	Builds the sortie requirements by MDS/sortie/pilot class from *tblRapSortieReqmts* and *tblRapSortieDistribution*.

FORMS

Forms are used to provide an improved graphic interface on the computer screen, facilitating the display or entry of data. Since data can be entered directly into a table or displayed in the form of a table or query result, forms are not essential to use of the database. However, they can be very useful in making the database accessible to nonexpert users or to control input of data from outside sources.

REPORTS

Reports are similar to forms but are designed to produce a user-friendly display of data on the printed page rather than the computer screen. No standard reports have been developed for the range and airspace database.

MACROS AND MODULES

Macros are routines saved by users to facilitate the execution of recurring tasks. Modules are collections of Visual Basic code saved by users for execution of recurring tasks. This project has used modules for various functions involved in the construction of the database. The module file *sptQryAssessRanges*, for instance, contains functions that are directly evaluated in the process of comparing infrastructure with requirements.

NAMING CONVENTIONS

Although *Access* has liberal rules for naming tables, queries, forms, and fields within it, we adopted somewhat more stringent conventions for naming the essential components of the range and airspace database. We anticipated that development of some database features might require Visual Basic programming or manually authored SQL statements. To facilitate such programming, we adopted a convention of avoiding blank characters within names of the standard objects. Likewise, for field labels, we decided to avoid blank characters.

To have the name indicate the object type, we attach a prefix indicating the object type: *tbl* for tables and *frm* for forms. For queries, we use one of several prefixes to denote the query type. These are:

qapp for an append query

qdel for a delete query

qmak for a make-table query

qsel for a select query

quni for a union query

qupd for an update query.

If a table, query, or form does not have one of these prefixes in its name, it is generally an ad hoc retrieval or display of information. Several of these are provided as examples of how the database can be used.

Our discipline in adhering to these conventions was not perfect. Occasional violations of our naming conventions do not impair the functionality of the database or prevent use of the affected names in programming; they simply make their use less convenient.

Chapter Five

A WEB "TOUR" OF THE SYSTEM

The database and its web interface were designed to be installed on a server at Hq ACC with a hyperlink from the ACC/DOR home page. This chapter, designed to be used by a reader with access to the database via a web browser, takes the reader through a brief tour of the system.

BASIC INQUIRIES

Figure 5.1 displays the top page which, in turn, leads to the rest of the web DSS.

Clicking on the hyperlink at the bottom of the top page leads to the system's table of contents, as shown in Figure 5.2.

The table of contents page shows the user some of the logic of the embedded models:

- Starts with the joint mission framework
- Presents a definition of sortie types and infrastructure resource categories (e.g., ranges, routes, etc.)
- Links the JMF to sortie requirements, then shows, for each infrastructure resource category, the supporting infrastructure characteristics required for each type of sortie
- Presents the characteristics of existing infrastructure

RAND MR1286/1-5.1

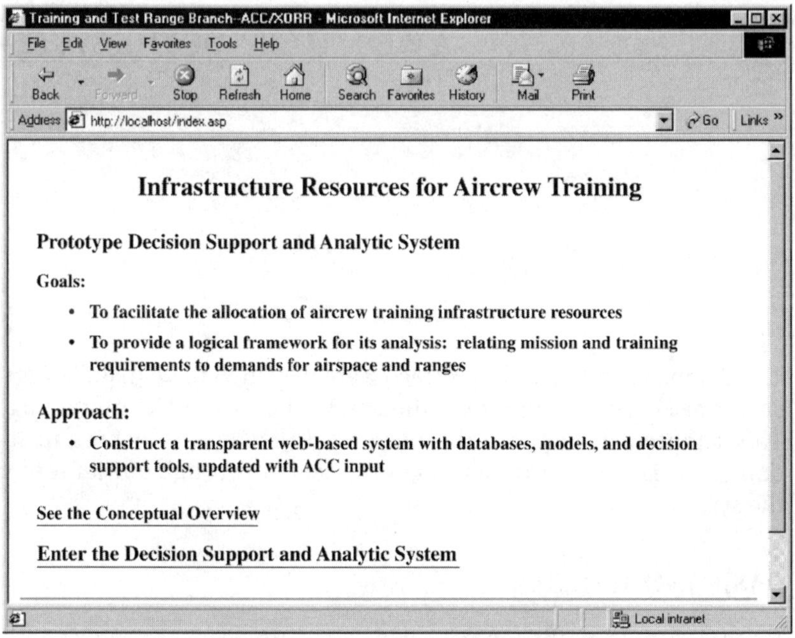

Training and Test Range Branch–ACC/XORR - Microsoft Internet Explorer

File Edit View Favorites Tools Help

Back Forward Stop Refresh Home Search Favorites History Mail Print

Address http://localhost/index.asp Go Links

Infrastructure Resources for Aircrew Training

Prototype Decision Support and Analytic System

Goals:

- **To facilitate the allocation of aircrew training infrastructure resources**
- **To provide a logical framework for its analysis: relating mission and training requirements to demands for airspace and ranges**

Approach:

- **Construct a transparent web-based system with databases, models, and decision support tools, updated with ACC input**

See the Conceptual Overview

Enter the Decision Support and Analytic System

Local intranet

Figure 5.1—Top-Level Web Page

- Compares the requirements and availabilities for user-specified sorties and infrastructure resources (a color-coded display summarizes the results).

In addition to generating lists of useful information (e.g., MDS characteristics or range location displayed in a map of the United States), the DSS can identify infrastructure resources within a user-specified distance from a particular base. More important, it can show whether a certain infrastructure resource—say, a particular range—has the needed characteristics for specified sortie types and related training needs (e.g., whether a specified range has laser scoring capability).

RAND *MR1286/1-5.2*

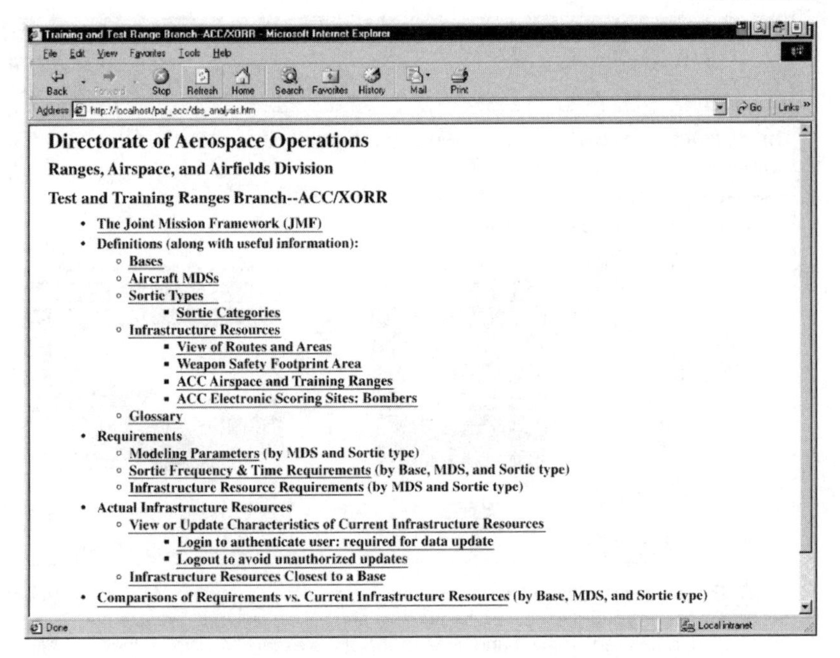

Figure 5.2—Web Page with Table of Contents

ADVANCED USE OF THE DSS

At a more advanced level, color-coded displays generated by the DSS can be used to compare the actual characteristics of an infrastructure resource, say, a range, with the requirements of a particular MDS/ sortie-type combination (see Figure 5.3).

Providing local range and airspace managers a convenient process to update the characteristics of existing infrastructure resources is key to maintaining current information on existing assets. The range and airspace DSS provides this capability. The page used to invoke the update process is shown as Figure 5.4. Software in the DSS checks to ensure that updates to the database are accepted only from

RAND *MR1286/1-5.3*

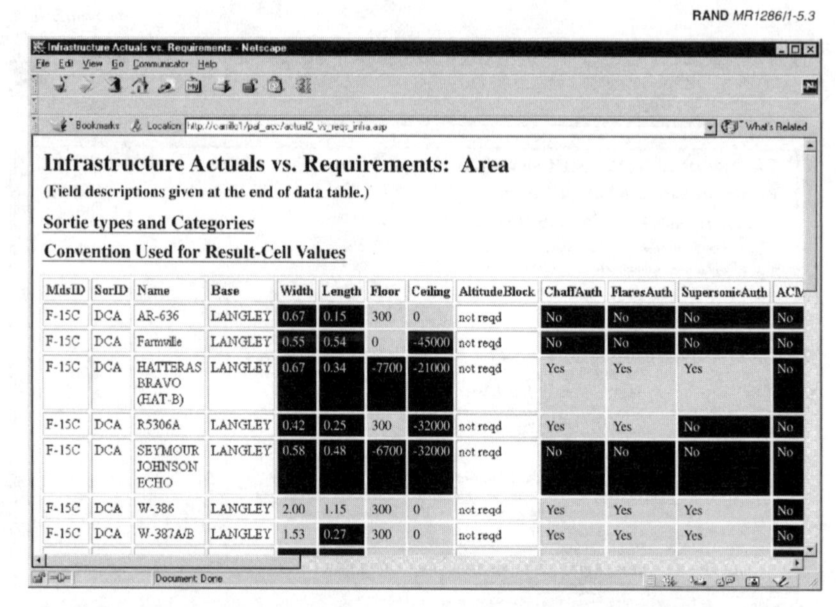

Infrastructure Actuals vs. Requirements: Area

(Field descriptions given at the end of data table.)

Sortie types and Categories

Convention Used for Result-Cell Values

MdsID	SorID	Name	Base	Width	Length	Floor	Ceiling	AltitudeBlock	ChaffAuth	FlaresAuth	SupersonicAuth	ACM
F-15C	DCA	AR-636	LANGLEY	0.67	0.15	300	0	not reqd	No	No	No	No
F-15C	DCA	Farmville	LANGLEY	0.55	0.54	0	-45000	not reqd	No	No	No	No
F-15C	DCA	HATTERAS BRAVO (HAT-B)	LANGLEY	0.67	0.34	-7700	-21000	not reqd	Yes	Yes	Yes	No
F-15C	DCA	R5306A	LANGLEY	0.42	0.25	300	-32000	not reqd	Yes	Yes	No	No
F-15C	DCA	SEYMOUR JOHNSON ECHO	LANGLEY	0.58	0.48	-6700	-32000	not reqd	No	No	No	No
F-15C	DCA	W-386	LANGLEY	2.00	1.15	300	0	not reqd	Yes	Yes	Yes	No
F-15C	DCA	W-387A/B	LANGLEY	1.53	0.27	300	0	not reqd	Yes	Yes	Yes	No

NOTE: Entries in width and length columns indicate the proportion of the requirement met by dimensions of the asset. Entries in floor and ceiling columns indicate the difference between requirement and dimensions of the asset. "Yes/No" entries in other columns indicate whether or not required characteristic is available on the asset.

Figure 5.3—Web Page Comparing Current Assets with Requirements

specified sources. The page used for this verification is shown as Figure 5.5. As a further protection against erroneous updates, a system administrator must validate the new data entries before master copies of database tables are updated. Additionally, the system administrator must run models offline to update the results provided in color-coded displays.

RAND *MR1286/1-5.4*

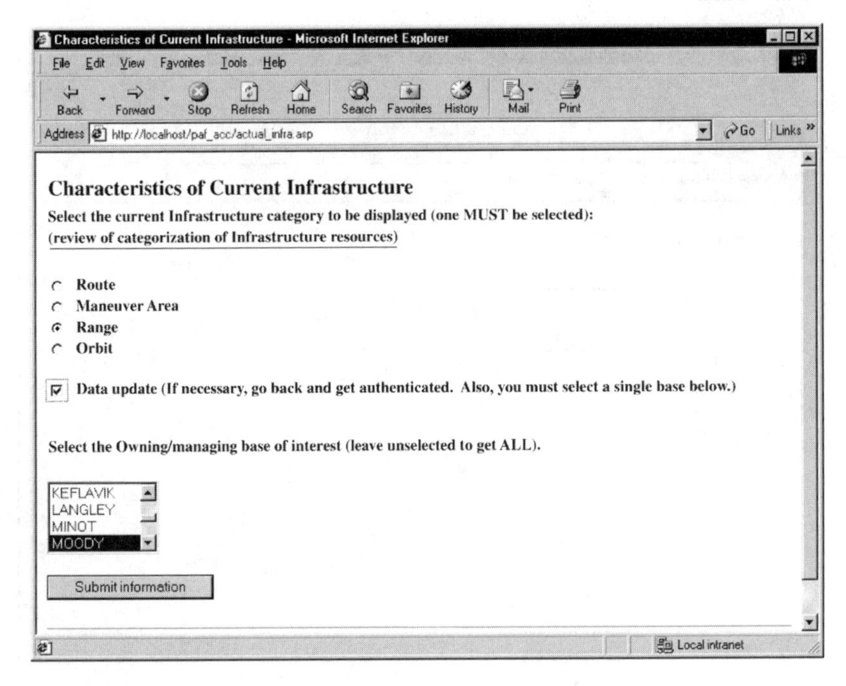

Figure 5.4—Web Page to Update Infrastructure Characteristics

RAND *MR1286/1-5.5*

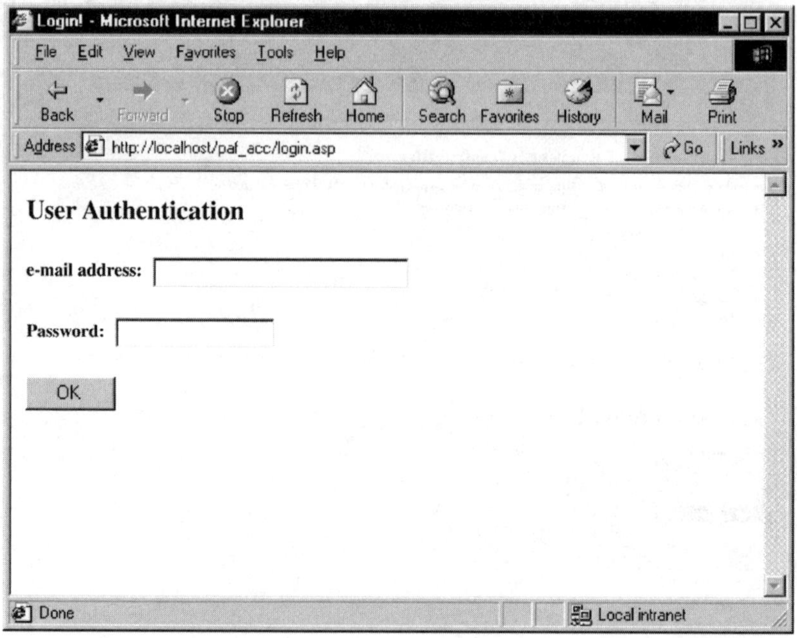

Figure 5.5—Web Page to Authenticate a User for Database Update

Chapter Six

SYSTEM MAINTENANCE AND DEVELOPMENT

This chapter discusses both ongoing requirements for maintenance of the range and airspace DSS and the need for additional development. Additional development is required both to eliminate currently identified system shortcomings and to add useful functionality.

KEEPING THE SYSTEM VIABLE

The range and airspace DSS provides a powerful tool for range and airspace managers and a potential tool for other aircrew training resource managers. The system requires an administrator with the following competencies:

- Database update, retrieval, and development
- Visual Basic programming
- Interfaces among databases, web servers, and web browsers
- Familiarity with aircrew continuation training requirements
- Familiarity with range and airspace infrastructure characteristics.

Additionally, other range and airspace managers must become familiar with the DSS's contents and capabilities so that they can place appropriate demands upon it. Also, as motivation to keep the system updated, data sources must perceive that the DSS is used advantageously in addressing critical issues. It is axiomatic that a system perceived to be unused will also be poorly maintained.

DATA MAINTENANCE ISSUES

Almost all the data in the DSS are subject to change and therefore need to be maintained.

Joint Mission Framework

The JMF does not change rapidly at the strategic level, but one can expect that as weapon systems evolve and as experience is gained in theaters such as Kosovo, the match of operational tasks to objectives will evolve.[1] Several changes per year at this level of the JMF can be expected. Additionally, the database will have to be changed to reflect modifications to the linkages between operational tasks and training sorties. This is probably best done centrally and at least annually.

Training Requirements

Training requirements will change, as indicated above, owing to the changing nature of Air Force doctrine as it responds to joint force requirements, and the database must be updated to reflect these changes. RAP tasking messages can serve to identify new and changed demands for sortie types. Additionally, the infrastructure requirements associated with various sortie types will change as weapon systems and tactics evolve. Capturing this changing data, which should probably be done at least annually, will require access to sources familiar with the needs and objectives of various sortie types.

The demand for training infrastructure depends on the product of the training requirements per pilot and the number of pilots needing training. Information about both the number of pilots (classified by MDS, experienced vs. inexperienced, and CMR vs. BMC) and their location will also change with time and should be updated at least annually.

---

[1]For example, during the project the number of joint missions was expanded from six to eleven to accommodate new families of operations other than war that U.S. Air Forces in Europe (USAFE) found important.

As currently developed, the range and airspace DSS does not recognize practical constraints on the number of sorties achievable at each location. The total number of available sorties may be less than the required sorties used in our demand calculations.

Infrastructure Requirements

Maintenance of data on current infrastructure is best done via a web interface, where the local managers of infrastructure assets would update information about the assets used by their local aircrews. A web interface seems to be the most efficient approach to this maintenance task. We have provided a rudimentary web interface for this purpose in the DSS, but operational procedures for its use must be developed and implemented by ACC. For example, where assets are used by multiple bases but managed by none of them (e.g, a Navy-managed asset), responsibility for updates must be assigned.

Comparing Requirements with Current Infrastructure

The DSS contains database tables and web-accessible displays that depict whether current infrastructure assets meet the requirements for various MDS/sortie-type combinations. The DSS contains modules that automate the update of these tables and displays, but the updates are not fully automatic. Whenever requirements or current asset information is changed, a system administrator must execute a series of routines to update the comparison tables.

NEW DEVELOPMENT

Assessing Geographical, Qualitative, and Quantitative Factors Simultaneously

In the companion document, *Relating Ranges and Airspace to Air Combat Command Mission and Training Requirements*, MR-1286-AF, we assessed current range and airspace in three ways—geographically, qualitatively, and quantitatively. Each of these assessments was performed independently. For example, when evaluating whether current assets provide sufficient capacity (the quantitative assessment), we considered all assets currently used by each unit, including some that are beyond ideal geographic limits. Similarly, the analyses reveal that some units can avail themselves of higher-quality assets but must travel longer distances (reducing training event durations) to reach them.[2]

The analytical capabilities of the DSS would be strengthened if geographical, qualitative, and quantitative factors could be considered simultaneously, so that explicit tradeoffs among the factors could be more readily visualized. This can be done using operations research methods, such as linear programming.

Evaluation of Other Training Resources

With some additional investment, the DSS could be expanded to permit efficient calculations of requirements for other training resources, such as flying hours and munitions. Consumption factors, such as average sortie duration or munitions consumed per sortie, would be required at the appropriate level of detail (presumably base/MDS/sortie). With these additions to the database, ACC/DOR could develop training resource requirements in a more consistent and balanced way and provide more specific justification to planners and programmers.

Other Range and Airspace Management Issues

ACC/DOR could expand the database to include other management information that must be exchanged routinely between headquarters and field units. Some examples might include:

- Detailed inventories of scoring, target, and threat emitter equipment
- Range and airspace utilization data
- Weapon safety footprint data
- Environmental impact data

---

[2]Use of the Townsend Range by Moody AFB units is a prime example.

- Data regarding land withdrawals, Federal Aviation Administration agreements, and other intergovernmental asset management arrangements.

Non-ACC Users

The DSS can be expanded to include requirements and infrastructure from non-ACC range and airspace users, possibly reserve components, Air Education and Training Command, Air Force Materiel Command, and other services.

Appendix A

MISSION/SORTIE DEFINITIONS USED IN THE DATABASE

Appendix A

Mission/Sortie Definitions Used in the Database

Group	Sortie ID	Full Name	Category[a]	Description
Basic flying, non-RAP	INS	Instrument	Basic	Sortie designed to achieve proficiency in instrument flying.
	AHC	Aircraft handling characteristics	Basic	Training for proficiency in utilization and exploitation of the aircraft flight envelope, consistent with operational and safety constraints, including but not limited to high/maximum angle of attack maneuvering, energy management, minimum time turns, maximum/optimum acceleration and deceleration techniques, and confidence maneuvers. Sortie may also include basic aircraft navigation and instrument approaches. May be referred to as a Defensive Tactics (DT) sortie in bomber contexts.
	CP	Crew proficiency	Basic	Sortie designed to achieve proficiency in basic flying skills; used in aircraft types with nonaircrew mission crew positions.
	CON	Contact	Basic	Sortie designed to achieve proficiency in helicopter takeoff and landing patterns.
Basic Combat	BFM	Basic fighter maneuver	Basic	Sortie designed to apply aircraft handling skills to gain proficiency in recognizing and solving range, closure, aspect, angle off, and turning room problems in relation to another aircraft to attain a position from which weapons may be launched or to defeat weapons employed by an adversary. Scale is 1 v. 1. Maneuvers are within visual range.
	ACM	Air combat maneuver	Basic	Sortie designed to achieve proficiency in element formation maneuvering and the coordinated application of BFM to achieve a simulated kill or effectively defend against one or more aircraft from a preplanned starting position. For range and airspace requirements, scale is assumed to be 2 v..X Maneuvers are generally within visual range.

Appendix A—continued

Group	Sortie ID	Full Name	Category[a]	Description
Basic Combat	BSA	Basic surface attack	Basic	Sortie designed to achieve proficiency in medium/low altitude tactical navigation and air-to-surface weapon delivery events.
	CSS	Combat skills sortie	Basic	Sortie designed to achieve proficiency in selected events; events are accomplished independently (as building blocks) rather than being integrated in a specific combat scenario.
Fighter, air-to-air	DCA	Defensive counter air	Applied	Sortie designed to develop proficiency in defensive counter air tactics. For the purpose of determining range and airspace requirements, the sortie is assumed to be on a 4 v. X scale. Full vertical dimension is used to accomplish tactical objectives.
	OCA	Offensive counter air	Applied	Sortie designed to develop proficiency in air-to-air offensive counter air tactics. For the purpose of determining range and airspace requirements, the sortie is assumed to be on a 4 v. X scale. Full vertical dimension is used to accomplish tactical objectives.
	OCA ANTI-HELO	Anti-helicopter mission	Variant	A-10 mission to gain proficiency in attacks against opposing helicopters.
Fighter, air-to-ground	SAT FTR	Surface attack tactics (fighter)	Applied	Sortie designed to develop proficiency in surface attack against a tactical target; should include air or ground threat. Although not specified in AFI 11-2 series publications, the sortie should include tactical navigation events during ingress and egress. For the F-117, includes vertical navigation events.
	SAT FTR GD OPP	SAT (fighter) ground threat opposed	Variant	SAT variant using ground threat emitters to provide simulated opposition.
	SAT FTR AIR OPP	SAT (fighter) air threat Opposed	Variant	SAT variant using Red Air assets to provide simulated opposition.
	SAT FTR LIVE	SAT (fighter) with live ordnance	Variant	SAT sortie with delivery of live weapons.

Appendix A—continued

Group	Sortie ID	Full Name	Category[a]	Description
Fighter, air-to-ground	SAT FTR CLASS OPS	SAT (fighter) with classified operations	Variant	F-117 variant that includes operations to be performed on a classified range.
	SEAD-C	Suppression of enemy air defense, conventional	Applied	Sortie designed to develop proficiency in suppression of enemy air defenses using conventional air-to-ground weapons.
	SEAD	Suppression of enemy air defense	Applied	Sortie designed to develop proficiency in suppression of enemy air defenses using antiradiation weapons.
	FAC-A	Forward air control	Applied	Sortie flown to provide airborne forward air control of armed attack fighters in support of actual or simulated ground fighters.
Bomber	SAT BOMB	Surface attack tactics (bomber)	Applied	Sortie designed to develop proficiency in surface attacks against a tactical target; should include air or ground threat. Sortie should include tactical navigation events during ingress and egress.
	SAT BOMB INRT LO	SAT (bomber) low altitude	Variant	SAT BOMB variant in which actual release of inert ordnance occurs from low altitude.
	SAT BOMB INRT HI/MED	SAT (bomber) hi/medium altitude	Variant	SAT BOMB variant in which actual release of inert ordnance occurs from medium or high altitude.
	SAT BOMB LIVE	SAT (bomber) with live ordnance	Variant	SAT sortie with delivery of live weapons.
	SAT BOMB SIM	SAT (bomber) simulated release	Variant	SAT BOMB variant in which release of weapons is simulated.

Appendix A—continued

Group	Sortie ID	Full Name	Category[a]	Description
Bomber	SAT BOMB MARITIME	SAT (bomber) maritime	Variant	SAT BOMB variant flown to practice mining a maritime target.
Other	MSN HC130	Mission, HC-130	Applied	Combat scenario profile that relates to the requirements of the unit's designed operational capability (DOC) statement.
	MSN HC130 WATER	Mission, HC-130 over water	Variant	HC-130 search-and-rescue mission variant performed over water.
	MSN E8C	Mission, JSTARS	Applied	Combat scenario profile that relates to the requirements of the unit's DOC statement.
	MSN E8C TM	JSTARS mission with terrain masking	Variant	JSTARS mission sortie that includes terrain masking.
	MSN E8C RETRO	JSTARS mission with retrograde event	Variant	JSTARS mission sortie that includes a combat separation event (departure from orbit and rapid descent to gain airspeed).
	MSN E3	Mission, AWACS	Applied	Combat scenario profile that relates to the requirements of the unit's DOC statement.
	MSN EC130H	Mission, Compass Call	Applied	Combat scenario profile that relates to the requirements of the unit's DOC statement.
	MSN EC130E	Mission, ABCCC	Applied	Combat scenario profile that relates to the requirements of the unit's DOC statement.
	MSN RC135	Mission, Rivet Joint	Applied	Combat scenario profile that relates to the requirements of the unit's DOC statement.

Appendix A—continued

Group	Sortie ID	Full Name	Category[a]	Description
Other	MSN U2	Mission, U-2	Applied	Combat scenario profile that relates to the requirements of the unit's DOC statement.
	MSN HH60G	Mission, HH-60G	Applied	Combat scenario profile that relates to the requirements of the unit's DOC statement.
	MSN UAV	Mission, UAV	Applied	Combat scenario profile that relates to the requirements of the unit's DOC statement.
Combined	AWACS A-A	AWACS with air-to-air	Combined	Sortie designed to exercise AWACS with air-to-air fighters.
	AWACS EC	AWACS electronic combat	Combined	Sortie designed to exercise AWACS employment in conjunction with a SEAD mission.
	CSAR	Combat search and rescue	Combined	Small multi-MDS exercise (SMME) that combines HH-60 MSN, SAT (F-16, A-10, F-15E), HH-130 MSN, and/or UAV MSN roles.
	FP/SWEEP	Force protection/ sweep exercise	Combined	SMME that combines OCA (F-15C, F-16, F-15E, F-22) and SAT (F-15E, F-16, F-117, A-10, B-1, B-2, B-52) roles. Targets may be on or off range.
	LFE	Large force engagement	Combined	Flag or other large training exercise involving multiple flights of aircraft types in a variety of roles; simulates the scale and complexity of actual combat. Combines SAT, OCA, AWACS MSN, JSTARS MSN, UAV MSN, ABCCC MSN, CAS, CSAR, and/or Rivet Joint MSN roles.

[a]*Basic* sorties are building-block exercises that are used to train fundamental flying and operational skills. *Applied* sorties are intended to more realistically simulate combat operations, incorporating intelligence scenarios and threat reaction events. *Variants* are subdivisions of Ready Aircrew Program (RAP) sorties, constructed for this study, that differ significantly from each other in their infrastructure requirements. *Combined* sorties are structured to bring together several MDS, performing different operational roles, in a single training mission.

Appendix B

RANGE AND AIRSPACE CHARACTERISTICS

The following table lists the range and airspace characteristics captured for our analysis. As indicated, some information was captured only for MDS/sortie types, some only for available assets (existing ranges and airspace), and some for both requirements and available assets. Some characteristics are in text form (e.g., scheduling agency), some are in numerical form (e.g., length in nautical miles), and some are in binary (yes/no) form (e.g., authorization to dispense chaff). Binary characteristics are punctuated using a question mark in the characteristics column. Binary characteristics are interpreted as indicating if the item is required (in a requirements array) or authorized (in an available assets array).

Threats are listed as a separate infrastructure type. However, threat emitters and communications jammers must be installed on a route, area, or range. In the database, threat requirements appear only once in any given MDS/sortie-type requirements array. However, threat infrastructure availability is recorded for each route, maneuver area, and range.

Appendix B

Range and Airspace Characteristics

Infrastructure Type	Characteristics	Require-ments	Available Assets
Low-level routes	Name/designation		X
	Reporting agency		X
	Scheduling agency		X
	Point of contact for scheduling agency		X
	Commercial phone for POC[a]		X
	DSN phone for POC		X
	Entry latitude (decimal degrees)		X
	Entry longitude (decimal degrees)		X
	Exit latitude (decimal degrees)		X
	Exit longitude (decimal degrees)		X
	Alternate entry points?		X
	Alternate exit points?		X
	Open 24 hours?		X
	Charted opening time		X
	Charted closing time		X
	Days per week		X
	Percentage of operating hours unavailable due to maintenance		X
	Percentage of operating hours used by non-ACC users		X
	Percentage of operating hours used by ACC users		X
	Flight spacing (minutes)		X
	Length (nm)		X
	Width (nm)		X
	Floor (ft above ground level [AGL])		X
	Ceiling (ft AGL)		X
	Route time (minutes)	X	
	Speed (knots)	X	
	Minimum width (nm)	X	
	Minimum length (nm)	X	
	Maximum floor (ft AGL)	X	
	Minimum ceiling (ft AGL)	X	
	Terrain-following operations?	X	X
	Segment below 300 ft?	X	X

Appendix B—continued

Infrastructure Type	Characteristics	Require-ments	Available Assets
	25 nm segment cleared up to 5000 ft?	X	X
	Instrument meteorological conditions (IMC)-capable?	X	X
	Percentage of route required to be mountainous	X	X
	Training route leads into/passes thru MOA or warning area?	X	X
	Name/designation of adjoining MOA or warning area		X
Maneuver areas	Name/designation		X
	Reporting agency		X
	Scheduling agency		X
	POC for scheduling agency		X
	Commercial phone for POC		X
	DSN phone for POC		X
	Latitude at center (decimal degrees)		X
	Longitude at center (decimal degrees)		X
	Open 24 hours?		X
	Charted opening time		X
	Charted closing time		X
	Days per week		X
	Percentage of operating hours unavailable due to maintenance		X
	Percentage of operating hours used by non-ACC users		X
	Percentage of operating hours used by ACC users		X
	Width		X
	Length		X
	Floor (ft)		X
	Floor type (AGL or mean sea level [MSL])		X
	Ceiling (ft MSL)		X
	Minimum width (nm)	X	
	Minimum length (nm)	X	
	Maximum floor (ft AGL)	X	
	Minimum ceiling (ft MSL)	X	

Appendix B—continued

Infrastructure Type	Characteristics	Require-ments	Available Assets
	Lowest floor for an altitude block (ft MSL)	X	
	Highest ceiling for an altitude block (ft MSL)	X	
	Minimum altitude block required (ft)	X	
	Chaff?	X	X
	Flares?	X	X
	Over land?	X	X
	Over water?	X	X
	Over mountains?	X	X
	Air-air communications?	X	X
	Air-ground communications?	X	X
	Datalink?	X	X
	Adjoining orbit?	X	X
	Name/designation of adjoining orbit		X
	Access to air-ground range?	X	X
	Name/designation of adjoining range		X
	ACMI?	X	X
	Supersonic operations?	X	X
Ranges	Name/designation		X
	Alternate name		X
	Complex		X
	Reporting agency		X
	Scheduling agency		X
	Scheduling base		X
	POC for scheduling agency		X
	Commercial phone for POC		X
	DSN phone for POC		X
	Latitude at center (decimal degrees)		X
	Longitude at center (decimal degrees)		X
	Open 24 hours?		X
	Charted opening time		X
	Charted closing time		X
	Days per week		X
	Percentage of operating hours unavailable due to maintenance		X

Appendix B—continued

Infrastructure Type	Characteristics	Require-ments	Available Assets
	Percentage of operating hours used by nonACC users		X
	Percentage of operating hours used by ACC users		X
	Width of restricted airspace (nm)		X
	Length of restricted airspace (nm)		X
	Ceiling of restricted airspace (ft MSL)		X
	Width of weapon safety footprint area		X
	Length of weapon safety footprint area		X
	Restricted airspace minimum width (nm)	X	
	Restricted airspace minimum length (nm)	X	
	Restricted airspace minimum ceiling (ft MSL)	X	
	Weapon safety footprint area minimum width (nm)	X	
	Weapon safety footprint area minimum length (nm)	X	
	Conventional circles?	X	X
	Strafe pits?	X	X
	Strafe targets 30mm authorized?	X	X
	Number of bomb targets scored simultaneously	X	X
	Lighted targets?	X	X
	Vertical targets?	X	X
	Tactical target array?	X	X
	Second tactical target array separated by 30 nm from the first array?	X	X
	Urban target array?	X	X
	Ordnance type (inert, live, or both)	X	X
	Number of laser targets required	X	X
	Number of infrared-significant (heated) targets	X	X
	Number of radar-significant targets	X	X
	Scoring no drop?	X	X
	Laser spot scoring?	X	X
	Night scoring?	X	X
	Scoring with 1-meter accuracy?	X	X

Appendix B—continued

Infrastructure Type	Characteristics	Require-ments	Available Assets
	Scoring available within 15 seconds of impact?	X	X
	Chaff/flare/ECM pods?	X	X
	Illumination flares?	X	X
	Attack heading variable by 90 degrees?	X	X
	Secured to allow classified operations?	X	X
	Night vision goggles?	X	X
	Part of range over water?	X	X
	Range control officer (RCO)?	X	X
Threats	Number of required threat emitters	X	X
	Multiple threat emitter?	X	X
	FSU[b] area defense emitter?	X	X
	Non-FSU threat emitter?	X	X
	Transportable threat emitter?	X	X
	Post-mission threat reaction debrief capability?	X	X
	Reactive threat emitter system?	X	X
	Smokey SAMs?	X	X
	Radar jammer?	X	X
	Communications jammer?	X	X
Orbits	Name/designation		X
	Type (refueling, mission)		X
	Reporting agency		X
	Scheduling agency		X
	POC for scheduling agency		X
	Commercial phone for POC		X
	DSN phone for POC		X
	Entry latitude (decimal degrees)		X
	Entry longitude (decimal degrees)		X
	Exit latitude (decimal degrees)		X
	Exit longitude (decimal degrees)		X
	Open 24 hours?		X
	Charted opening time		X
	Charted closing time		X

Appendix B—continued

Infrastructure Type	Characteristics	Require-ments	Available Assets
	Days per week		X
	Percentage of operating hours unavailable due to maintenance		X
	Percentage of operating hours used by non-ACC users		X
	Percentage of operating hours used by ACC users		X
	Length (nm)		X
	Width (nm)		X
	Floor (ft MSL)		X
	Ceiling (ft AGL)		X
	Minimum width for refueling (nm)	X	
	Minimum length for refueling	X	
	Maximum floor for refueling (ft MSL)	X	
	Minimum ceiling for refueling (ft)	X	
	Minimum floor for refueling altitude block (ft)	X	
	Maximum ceiling for refueling altitude block (ft)	X	
	Altitude block required for refueling (ft)	X	
	Minimum width for mission (nm)	X	
	Minimum length for mission (nm)	X	
	Maximum floor for mission (ft MSL)	X	
	Minimum ceiling for mission (ft)	X	
	Altitude block required for mission (ft)	X	
	Percentage of the orbit/track over mountainous terrain	X	X
	Radiatable air-to-ground or artillery range at 90–150 nm from orbit?	X	X
	Direct access to Army maneuver area or air-to-ground range?	X	X
	Air-to-air range 60–120 nm from orbit?	X	X
	Dedicated air-to-air frequency?	X	X
	Dedicated air-ground frequency?	X	X
	ABCCC training capsule?	X	X
	JTIDS datalink needed?	X	X
	Surveillance control datalink (SCDL)?	X	X

Appendix B—continued

Infrastructure Type	Characteristics	Require-ments	Available Assets
	Communications system operator training (CSOT) capability?	X	X
	JSTARS workstation?	X	X
Other	Threat air-to-air fighter?	X	
	Any air-to-air fighter?	X	
	Any air-to-ground fighter?	X	
	Threat air-to-ground fighters?	X	
	Heavy bomber?	X	
	Tanker?	X	
	E-3 (AWACS)?	X	
	E-8C (JSTARS)?	X	
	EC-130H (ABCCC)?	X	
	Ground FAC?	X	
	Ground control intercept?	X	
	Ground movers?	X	
	Post-mission truth data?	X	
	Landing zone?	X	

[a]Point of contact.

[b]Former Soviet Union.

Appendix C

DATA LIMITATIONS

This appendix addresses known limitations in the data used for our analysis and embedded in the range and airspace database. Limitations exist in data regarding both requirements and current infrastructure.

REQUIREMENTS-RELATED DATA PROBLEMS

Sortie Requirements per Pilot

Sortie requirements per pilot, used to calculate required infrastructure capacities, were derived primarily from annual RAP tasking messages. However, RAP messages do not include demands for basic skill sorties, such as AHC. We have largely ignored the demand for such sorties, assuming that these skills are practiced during sorties that are logged in other ways.

A more important problem is that the allocation of commander option sorties is subjective. We distributed commander option sorties to specified sortie types in proportion to how those types are represented in RAP tasking messages.

The sortie definitions used for this project do not correspond precisely to RAP sorties. First, we found it necessary to subdivide certain RAP sorties that have different infrastructure requirements, depending on how the sortie is flown. For example, we split SAT sorties into air- and ground-opposed variants, which have significantly different airspace requirements. These variants are just finer subdivisions of RAP sorties. We used our best judgment (assisted by ACC) to estimate the distribution of RAP sortie requirements among these variants.

Additionally, some sorties in our framework "collect" from multiple RAP sorties. These are designated *small multiple-MDS exercise* (SMME) sorties. They generally do not appear in RAP but were included in our framework to illustrate both the need for such sorties and to capture the additional infrastructure requirements they would entail. An example of this is AWACS AA, which trains interactions of air-to-air fighters with the Airborne Warning and Control System (AWACS). This sortie includes a fraction of DCA, SAT, and OCA sorties, but it has extra infrastructure requirements because of the need for an adjacent AWACS orbit that is properly oriented with the attack axis. Again, we had to use our best judgment to guess what fraction of each sortie type should train with AWACS because this requirement is not in RAP for AWACS or fighter combat crew members.

Number of Pilots

Pilot counts, also used to compute infrastructure capacity requirements, were based on PMAI and crew ratio data rather than actual head counts. This is appropriate because requirements based on the product of a base's PMAI and the MDS's crew ratio should provide a better (and more stable) average estimate of the demand for training infrastructure near a given base than would a head count. However, the PMAI data we used are subject to change and the crew ratio values we used are from multiple sources of varying reliability. Additionally, sortie counts depend on pilot experience levels, which are currently declining. The data for experience levels are based on a recent snapshot of pilot inventory. Similarly, RPI 6 pilots add an additional demand for training; their count is based on a snapshot of the actual inventory. In general, the dates for data for PMAI, crew ratios, experience levels, and RPI 6 are not the same.

Adjustments to Sortie Requirements

In computing the time demand for ranges and airspace, we inflated RAP-derived sortie counts to account for attrition (maintenance and weather cancellations), scheduling inefficiency, and noncontinuation training sorties (see discussion in Chapter Two). We found no empirical data from which to estimate these factors. The factors currently embedded in the range and airspace database should be reviewed and refined, if possible.

Peaks in Demand

Demand is not uniform over a year but can vary in response to phenomena such as preparation for and recovery from deployment. We assumed level demand throughout the year. However, to maintain appropriate readiness levels, sizing infrastructure supply to service such peaks might be more appropriate than sizing to average demands.

CURRENT INFRASTRUCTURE DATA PROBLEMS

Data describing currently available infrastructure have various problems. Data were collected using a spreadsheet form distributed throughout ACC. This discussion will be limited to data problems that were inherent in the forms (as opposed to problems with the responses) and which as a result limit the analysis that can be performed.

- Certain sorties, such as BFM and AHC, require block altitudes, whereas most sorties require a specific altitude range. However, the actual special-use airspace (SUA) floors are specified either as mean sea level (MSL) or above ground level (AGL), with the latter being unsuitable for the evaluation of block altitudes because ceilings are always specified as MSL. In general, both MSL and AGL floors should be provided, or (preferably) either one of these plus an average or maximum SUA ground altitude. Using the average would ignore the effect of widely varying altitudes over the SUA, whereas using the maximum might depict usable altitudes too conservatively.
- Opening/closing times and days per week are not as simple as the form would lead one to presuppose. One issue is how to treat cases in which reported opening/closing times are "sunrise/sunset." We replaced sunrise and sunset with 0600 and 1800, respectively, which is reasonable for training requirements spread across an annual cycle. However, seasonal variations in

training schedules (caused, for example, by contingency deployments) could make our assumption invalid. Also, when longer hours/days can be prearranged, the instructions ought to indicate that the longest workable period be indicated. The information of interest is not how much the infrastructure *is* open, but how much it *could be* open to satisfy demand. Even here, workload or funding limitations would presumably limit the maximum average period to something less than the maximum short-term open period.

- Yet another problem with opening and closing times is associated with infrastructure (especially military routes) that span time zones. Specifying zulu times in all cases is probably best.
- For routes, alternate entry and exit points are not currently usable in the range and airspace database because coordinate information is not supplied.

Composite Ranges and Areas

Ranges and airspace are often designated in sets. Elements of a set may be used individually or combined with contiguous elements of the set to produce ranges and SUA with greater lateral or vertical dimensions. These composite ranges and airspace are useful for training in the more space-demanding scenarios. When infrastructure is locally scarce, it is important to avoid double-counting the availability of an individual area that is part of a composite area. Most of the data supplied to us pertain to individual areas, but this is not always the case. For instance, Davis-Monthan AFB reports a single range for the entire Goldwater complex, whereas Hill AFB reports six separate range elements within the Utah Test and Training Range, designated by the restricted airspaces that cover them.

We have defined some composite areas, but our work is based on map data and may not account for unique situations that make it difficult to actually train in a composite area. For instance, some elements may be under different scheduling or air traffic control authorities.

Appendix D

HARDWARE AND SOFTWARE REQUIREMENTS

HARDWARE

There are no restrictions on the hardware of the web client other than speed requirements specified by the user.

The web server and the computer where the Access database is maintained have only one requirement: The hardware must be able to run the Microsoft web server products and to have sufficient speed to accommodate the projected number of users.

SOFTWARE

Web Server

The range and airspace DSS depends on the following Microsoft (MS) software installed on the web server:

- *Windows NT* (server or workstation) or Windows 95 or 98 operating system
- Personal Web Server (PWS) or Internet Information Server (IIS), version 4 or newer
- Microsoft Data Access Component (MDAC) version 2.

The DSS also uses a group of files located in a subdirectory on the path *inetsrv\iisadmin\website* of the *System* or *System32* directory of the operating system. The files contained in this subdirectory are

- *RangeAirspaceWeb.mdb* (an Access database)
- *usr_clave.mdb* (an Access database)
- *paf_acc.UDL* (an MS datalink file)
- paf_acc_w.UDL (an MS datalink file)
- *paf_acc_login.UDL* (an MS datalink file).

These files are distributed with the DSS; however, the properties of UDL files (location of database, in particular) must be modified for local application using standard operating system methods.

Finally, the DSS uses a group of files located in the web server subdirectory on the path *inetpub\wwwroot\paf_acc* or wherever that comparable directory is located at the time of installation of the DSS on a web server. This is the subdirectory where the Hypertext Markup Language (HTML) or Active Server Pages (ASP) files reside. These files are also distributed with the DSS.

HTML editors (such as MS Frontpage Express—part of MS Internet Explorer) may be used to modify *.HTM files. However, the MS Notepad editor should be sufficient.

Web Client

Web access to the range and airspace DSS depends on any of the major web browsers, such as MS Internet Explorer or Netscape version 4 or newer.

Web Database Update, Maintenance, and Queries

Microsoft Access is the application required to update databases, maintain or further develop the database, or perform queries. MS Excel may also prove useful in either preprocessing data provided to the database or postprocessing data retrieved from it.